21st Century Junior Library

DISCOVER THE COMPSOGNATHUS

Jennifer Zeiger

Our Prehistoric World: Dinosaurs

Published in the United States of America by:

Cherry Lake Press
2395 South Huron Parkway, Suite 200, Ann Arbor, Michigan 48104
www.cherrylakepress.com

Content Adviser: Gregory M. Erickson, PhD, Dinosaur Paleontologist, Department of Biological Science, Florida State University, Tallahassee, Florida

Reading Adviser: Beth Walker Gambro, MS, Ed., Reading Consultant, Yorkville, IL

Photo and Illustration Credits: © Stocktrek Images, Inc./Alamy, cover, title page; © Dotted Yeti/Shutterstock, 5; © Linda Bucklin/Shutterstock, 6; Jim, the Photographer from Springfield PA, United States of America, CC BY 2.0, via Wikimedia Commons, 7; © Linda Bucklin/Shutterstock, 8-9; © Linda Bucklin/Shutterstock, 10; © Dotted Yeti/Shutterstock, 13; © STT0007399/Media Bakery, 14; © STT0007398/Media Bakery, 16; © Dotted Yeti/Shutterstock, 19; © ASSOCIATED PRESS, 20

Cherry Lake Press is an imprint of Cherry Lake Publishing Group.

Library of Congress Cataloging-in-Publication Data has been filed and is available at catalog.loc.gov.

Cherry Lake Press would like to acknowledge the work of the Partnership for 21st Century Learning, a Network of Battelle for Kids. Please visit Battelle for Kids online for more information.

Printed in the United States of America

Cherry Lake Press

CONTENTS

WHAT WAS A COMPSOGNATHUS?

It is a warm, sunny afternoon. A giant **dinosaur** reaches its long neck into a tree to eat leaves. Another large dinosaur with big, sharp teeth sniffs the air. Far below them is a much smaller dinosaur. It darts after a lizard on the ground. This little dinosaur is known as the *Compsognathus*.

The *Compsognathus* was small but fast.

The *Compsognathus* lived between 161 million and 146 million years ago in what is now Europe. It was one of the smallest dinosaurs in history. The *Compsognathus* only grew to be 2 to 3 feet (.6 to .9 centimeters) long. It was only a little bigger than a chicken. Other dinosaurs alive at this time were as big as six elephants!

The *Compsognathus* was tiny compared to other dinosaurs, including the *Giganotosaurus*.

Think!

Most dinosaurs were much larger than the *Compsognathus*. How do you think the *Compsognathus*'s small size helped it? How did its size hurt it? Why do you think other dinosaurs were so big?

WHAT DID A COMPSOGNATHUS LOOK LIKE?

The *Compsognathus* stood on two legs. Its front arms were very small, but its back legs were long and very strong. They helped it run fast. The *Compsognathus* could run fast enough to catch a speedy lizard.

The *Compsognathus* weighed about the same as a house cat.

A *Compsognathus*'s strong legs helped it outrun other dinosaurs.

Birds have **hollow** bones. This makes them lightweight so they can fly. The *Compsognathus* also had hollow bones. It did not weigh much. The *Compsognathus* weighed about 12 pounds (5.4 kilograms). But the *Compsognathus* did not fly. Instead, its hollow bones helped it run very fast without using much energy.

The *Compsognathus* had a long tail. The tail helped it stay **balanced**. Balance allowed it to run and turn without falling.

The *Compsognathus* also had a long neck. The neck could bend easily. This helped the *Compsognathus* look around without turning its body.

Create!

Draw a picture of a *Compsognathus* on a piece of paper. Label its legs, arms, tail, and neck. Try to label parts of the body you haven't read about yet, too. How did each part help the dinosaur survive?

Balance allowed the *Compsognathus* to run fast and make quick turns.

Scientists know the *Compsognathus* ate lizards. They know this because the bones of a lizard were found in a *Compsognathus*'s stomach.

HOW DID A COMPSOGNATHUS LIVE?

A **carnivore** is an animal that eats meat. The *Compsognathus* was a carnivore. Scientists believe it ate lizards, bugs, and fish. It also ate small **mammals** similar to mice and rats. It used its speed and balance to catch them.

The *Compsognathus*'s name means "dainty jaw." It was named this because of its tiny, sharp teeth.

The *Compsognathus* had a long, **slender** head. Each of its two eyes was on either side of its head. These eyes were big. They gave the *Compsognathus* sharp vision.

Look!

Take a look at a picture of a *Compsognathus*. Check out its teeth and claws. Do they look scary? Would you want to run into a living *Compsognathus*? Does its small size change how dangerous it would be?

Each finger on the *Compsognathus*'s hands had a sharp claw. The *Compsognathus* used these claws to grab **prey** and pick them up. It also used sharp toe claws to pin prey to the ground. The *Compsognathus* could also pick up prey with its sharp teeth.

The *Compsognathus* had a small extra claw above its ankle. This claw is called a dewclaw.

Scientists learn more each time another *Compsognathus* fossil is found.

There are no *Compsognathuses* living today. Scientists study them by looking at **fossils**. They are still learning about the *Compsognathus*. Did it have feathers? Was it a good swimmer? Maybe one day you will be the scientist who discovers the truth!

GLOSSARY

balanced (BAL-unsd) steady, able to stay up without falling

carnivore (KAHR-nuh-vor) an animal that eats meat

dinosaur (DYE-nuh-sor) a kind of large reptile that lived millions of years ago

fossils (FAH-suhlz) the preserved remains of living things from thousands or millions of years ago

hollow (HAH-loh) empty inside

mammals (MAM-uhlz) one kind of animal that has fur and usually gives birth to live babies

prey (PRAY) an animal that is hunted by other animals for food

slender (SLEN-dur) thin

FIND OUT MORE

Books

Bingham, Caroline. *DK First Dinosaur Encyclopedia*. New York, NY: DK Publishing, 2007.

Gray, Susan H. *Compsognathus*. Chanhassen, MN: The Child's World, 2010.

Websites

With an adult, learn more online with these suggested searches.

Enchanted Learning—Compsognathus

Read all about the *Compsognathus* and print out a coloring page of the dino.

KidsDinos.com—Compsognathus

Learn more about the *Compsognathus* and follow links to learn about lots of other dinosaurs.

INDEX

ABOUT THE AUTHOR

Jennifer Zeiger graduated from DePaul University. She now lives in Chicago, Illinois, where she writes and edits books for children.